NATIONAL GEOGRAPHIC

School Publishing

Winning Properties

PATHFINDER EDITION

By Susan Halko

CONTENTS

Winning Properties

Mystery Matter

Here's a riddle: To play this game, you will need this object to score:

- It has a light brown color.
- It is thicker at one end than the other.
- It has a smooth texture.
- It has a mass of 56.7 grams (32 ounces).
- It is hard, not soft.
- When it hits a ball, you can hear a CRACK!

What is it?

You can use science to help you answer the riddle.

Matter is anything that takes up space. This book is matter. Your chair is matter. You are matter!

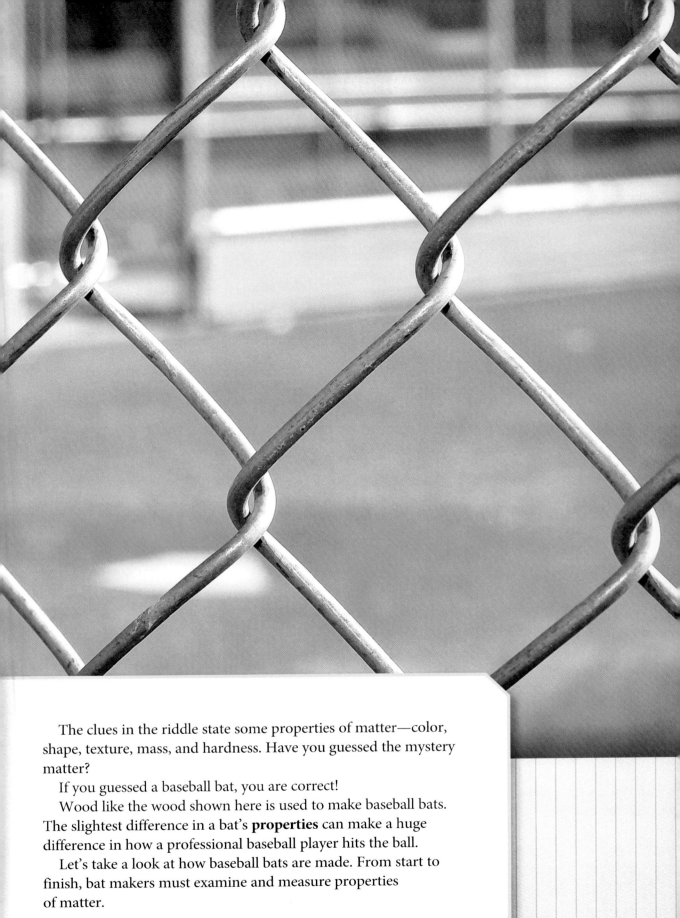

The clues in the riddle state some properties of matter—color, shape, texture, mass, and hardness. Have you guessed the mystery matter?

If you guessed a baseball bat, you are correct!

Wood like the wood shown here is used to make baseball bats. The slightest difference in a bat's **properties** can make a huge difference in how a professional baseball player hits the ball.

Let's take a look at how baseball bats are made. From start to finish, bat makers must examine and measure properties of matter.

Inspecting the Wood

Most bats you buy in a store are made out of a metal. But professional baseball players have to use bats made of all wood. Most players choose ash or maple. These types of wood are hard, strong, and not too heavy.

First, an inspector looks closely at the grains in the wood to make sure they are straight and evenly spaced. Crooked grains might lead to a weaker bat.

He also looks at the spacing of grains to check how dense it is, or how tightly its matter is packed together. The closer the grains are, the more dense it is. Baseball players want the grains to be dense for a stronger bat. The bat should not break when it hits a ball.

Next, the wood is cut into wedges. These wedges are put onto a machine called a lathe. The lathe spins the wood, and a worker uses a sharp tool to smooth the rough edges. The result is a rounded piece of wood, called a **billet**.

The billets are dried. This helps strengthen the wood. Then they are sorted by mass.

Spinning the Billets

The next step is for the billet to go on another lathe. This lathe has a computer. As the lathe spins the wood, it forms it into the shape of a bat. The computer runs the machine.

Measuring Mass. These billets are ready to have their masses measured.

Taking Shape. This lathe is spinning the billet. The parts of the bat are beginning to take shape.

knob   handle

The worker sands the bat again to get a smooth texture and to get it to the right mass. These bats are inspected and sorted.

Personalized Properties

The Major League Baseball Association has rules about the size of a bat. The thickest part—the barrel—can be no more than 6.9 centimeters (2.75 inches) thick. No bat can be longer than 106 centimeters (42 inches).

Even with these rules, baseball players can change the properties of their bats.

They can order a certain mass, length, and model. The model tells the exact shape the bat will take. One bat can have many slight changes in shape compared to another bat.

For example, one bat might have a thicker handle than another. The knobs can vary, too. Some are more narrow than others.

The Finishing Touch

Finally, the manufacturer's name is branded, or burned, onto the bat. Varnish is put on the bat. It's ready for the big leagues!

Sanding Down. A worker sands this bat. This shaves off wood to get it to the right mass.

Shiny Finish. The varnish helps give the bat its color and protects the wood.

barrel

end

Properties Matter

Professional ball players can be very choosy about the specific properties of their bats. Some players visit lumberyards to find the perfect piece of wood. Some players even watch the lathe operator make the bat to make sure it is just right.

Why do properties matter so much in a baseball bat? Let's see why mass and type of wood are important.

The Ideal Bat

Why do some players choose a bat with more mass and others choose a bat with less mass? In general, the less mass in the bat, the faster a player can swing it. On the other hand, a bat with more mass can hit a ball with more force.

So the ideal bat has enough mass to give the ball a solid hit but can still be swung fast. The perfect mass of a bat depends on an individual player's size, strength, and how fast he can swing.

Today, players generally use bats with less mass than in the past. Babe Ruth used a 1,134 gram (40 ounce) bat in 1927 when he hit 60 home runs. Most players today hit with an 850–936 gram (30–33 ounce) bat.

Pro's Picks

Louisville Slugger® is the official baseball bat of the Major League Baseball Association. The company has been making bats for over 100 years.

The company keeps track of the models ordered by professional ball players. Some pro players have let the company put their names on bats that match their specifications.

Louisville Slugger offers these models to players who want a bat just like their favorite pro player's bat. Check out these pros and their favorite bats!

Derek Jeter [New York Yankees]
Preferred bat: Black Smith Finish Louisville Slugger model P72
Length: 86 centimeters (34 inches)
Mass: 907 grams (32 ounces)
Wood: Ash

Jim Thome [Minnesota Twins, formerly Los Angeles Dodgers]
Preferred bat: Black Smith Finish Louisville Slugger model M356
Length: 87 centimeters (34.5 inches)
Mass: 907 grams (32 ounces)
Wood: Maple

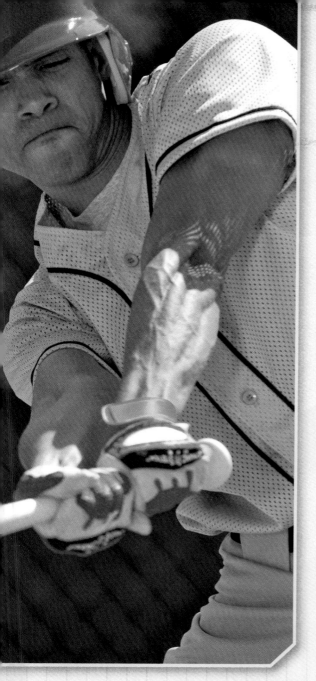

Maple or Ash?

Some players prefer bats made out of ash. Some choose maple. What's the difference?

Maple has more mass than ash. Its surface is harder, too. Many players like the harder surface of a maple bat.

Ash has less mass than maple. It is more flexible, too. Many players like an ash bat because they can have a bigger barrel. Some think a bigger barrel makes it easier to hit the ball.

The grains of ash and maple are different. The maple grains are tricky. An inspector might think he has a good piece of wood with straight grains but later finds weak spots that he couldn't see.

Ash grain   Maple grain

The structure of the grains in maple bats leads to a problem. When maple bats break, they don't just crack and fall apart. They send flying pieces of wood through the air. People have been hurt from these breaks.

Evan Longoria [Tampa Bay Rays]
Preferred bat: Heavy Flame Treated Louisville Slugger model I13
Length: 85 centimeters (33.5 inches)
Mass: 893 grams (31.5 ounces)
Wood: Ash

Research Challenge

Who is your favorite pro player? See if you can find out the properties of his favorite bat!

Wordwise

billet: piece of wood used to make a bat

matter: anything that has mass and takes up space

properties: something about an object that you can observe with your senses

Extreme Properties

A Big Hit. The Louisville Slugger Museum is in Louisville, Kentucky. It's connected to the factory where Louisville Slugger bats are made.

H&B Louisville Slugger Store

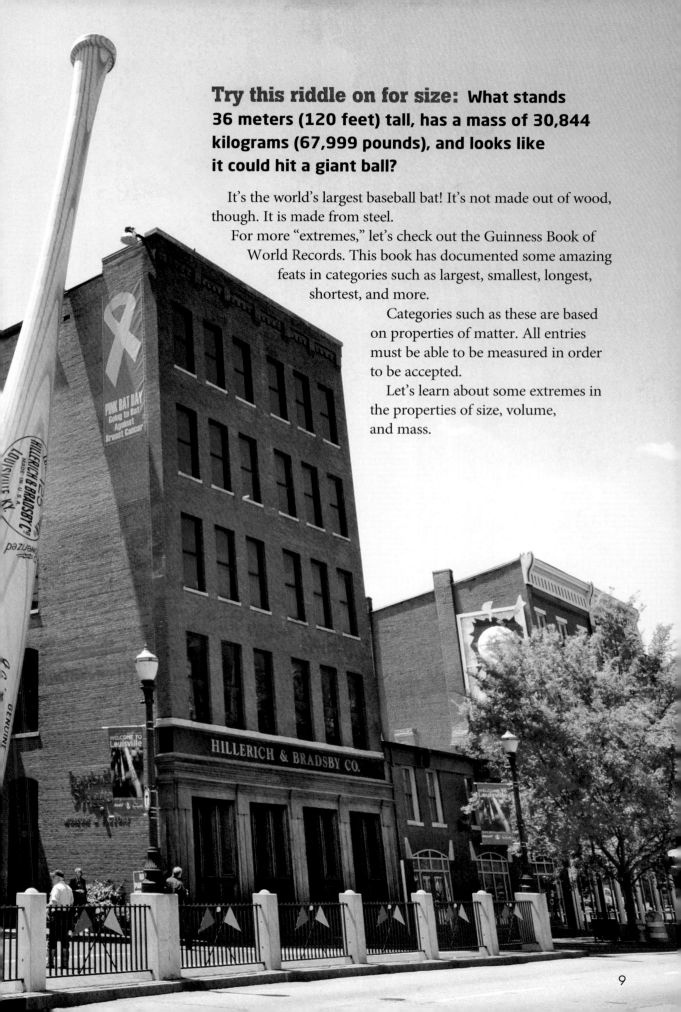

Try this riddle on for size: What stands 36 meters (120 feet) tall, has a mass of 30,844 kilograms (67,999 pounds), and looks like it could hit a giant ball?

It's the world's largest baseball bat! It's not made out of wood, though. It is made from steel.

For more "extremes," let's check out the Guinness Book of World Records. This book has documented some amazing feats in categories such as largest, smallest, longest, shortest, and more.

Categories such as these are based on properties of matter. All entries must be able to be measured in order to be accepted.

Let's learn about some extremes in the properties of size, volume, and mass.

Tallest Snowman

The world's tallest snowman measured 37 meters (122 feet) tall. That's taller than some office buildings! In February 2008, people from around Bethel, Maine, worked together to build this snow giant.

Its body was made from 5,897 metric tons (6,500 tons) of snow. Its eyelashes were made of 16 skis. Its mouth was made from painted tires.

Lots of Coffee

The world's largest coffee pot was made in Bosnia and measured 1.2 meters (4 feet, 1 inch) tall. That's about the average height of a fourth-grader! Its volume measured 800 liters (211 gallons). That's about 8,000 cups of coffee.

Everyday objects that are in the "largest" category must be made from the same materials as the regular-size object. Also, the large version must be able to work in the same way as the regular-size object. Just imagine the amount of coffee that comes out of this pot!

Big Dog, Little Dog

The mastiff and St. Bernard are the heaviest breeds of domestic dogs. Males of both types can have masses of more than 91 kilograms (200 pounds)! That's about 88 kilograms (195 pounds) heavier than the average Chihuahua. The Chihuahua is the world's smallest breed.

Mastiff

St. Bernard

Chihuahua

Properties Matter

Why do people measure properties of matter? Some people measure properties of matter to win a new world record. But most people measure properties of matter to learn more about the matter. It also helps them compare it to other matter—because matter is everywhere!

Describing and Measuring Matter

Let's check out what you learned about describing and measuring properties of matter.

1 What is matter?

2 How can looking at the grains in wood help you know if it is dense?

3 How does sanding a bat affect its mass?

4 What properties are important in a bat?

5 What kind of matter do you measure in your everyday life?